SILKWORMS

SILKWORMS

by Sylvia A. Johnson

Photographs by Isao Kishida

A Lerner Natural Science Book

Lerner Publications Company • Minneapolis

Sylvia A. Johnson, Series Editor

Translation of original text by Chaim Uri

Drawing on page 19 by Yoshitaka Moriue

The publisher wishes to thank Jerry W. Heaps, Department of Entomology, University of Minnesota, for his assistance in the preparation of this book.

The glossary on page 46 gives definitions and pronunciations of words shown in **bold type** in the text.

This book is available in two editions:
Library binding by Lerner Publications Company,
    a division of Lerner Publishing Group, Inc.
Soft cover by First Avenue Editions,
    an imprint of Lerner Publishing Group, Inc.
241 First Avenue North
Minneapolis, MN 55401 U.S.A.

Website address: www.lernerbooks.com

Library of Congress Cataloging-in-Publication Data

Johnson, Sylvia A.
  Silkworms.

  (A Lerner natural science book)
  Adaptation of: Kaiko/by Isao Kishida.
  Includes index.
  Summary: An introduction to the domesticated silkworm moth, raised on farms in Japan and elsewhere for the sake of the silk thread out of which its cocoons are constructed.
    1. Silkworms—Juvenile literature. [1. Silkworms. 2. Moths] I. Kishida, Isao, ill. II. Kishida, Isao. Kaiko. III. Title. IV. Series.
SF542.5.J63   638'.2   82-250
ISBN-13: 978-0-8225-1478-7 (lib. bdg. : alk. paper)
ISBN-10: 0-8225-1478-8 (lib. bdg. : alk. paper)
ISBN-13: 978-0-8225-9557-1 (pbk. : alk. paper)
ISBN-10: 0-8225-9557-5 (pbk. : alk. paper)
                    AACR2

International Standard Book Number: 0-8225-1478-8 (lib. bdg. : alk. paper)

                  0-8225-9557-5 (pbk. : alk. paper)
Library of Congress Catalog Number: 82-250

Manufactured in the United States of America
17 18 19 20 21 22 – JR – 13 12 11 10 09 08

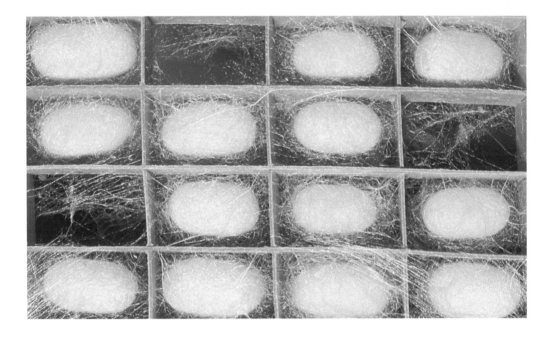

Take a close look at the photograph on this page. In it you can see a cardboard frame filled with small rectangular spaces. Each space contains a mysterious white object, oval in shape and about ½ inch (12.5 millimeters) long. The ovals are surrounded by networks of shiny, white threads.

What are these strange objects? Are they alive? Have they been made by humans or are they part of some natural process?

The objects in the frame are the **cocoons** of silkworm moths. From them come the silk fibers that people use to make strong, shiny silk cloth. The same cocoons that produce silk for human use are also part of the life cycle of the silkworm moth. In this book, you will find out how silkworm moths live and how people use the insects in producing silk.

The mulberry moth is one of the wild relatives of the silkworm moth.

The silkworm moth *(Bombyx mori)* is one of the thousands of species, or kinds, of moths that live throughout the world. Like their close relatives the butterflies, moths have wings covered with tiny scales and sense organs called **antennae** on their heads. Both moths and butterflies develop in a special way, going through four very different stages of growth as they become adults. This process of development is called **metamorphosis**, a combination of Greek words meaning "transformation."

The silkworm moth goes through the same development as other moths. But this insect is special because, unlike its relatives, it is **domesticated**. For centuries, silkworm moths have been raised and cared for by people. Today moths of this species are almost never found in the wild. Silkworm moths live on farms in Japan and other countries where silk is produced.

Like other domestic animals, silkworm moths have been affected by their long association with human beings. Their natural lives have been changed in some ways to make the insects more useful to their human owners.

An adult silkworm moth has a wingspan of about two inches (five centimeters). Its wings and body are covered with white scales.

The picture on the left shows male moths gathering around a small blue box. On the opposite page is a photograph taken from beneath the same box through a sheet of glass.

After they become adults, most moths live for only a short time. During these few days or weeks, the insects have only one important thing to do: they must produce more of their own kind before they die.

Like most other insects, moths reproduce sexually. Male and female moths must find partners of the opposite sex in order to mate and produce young. Moths living in the wild may have to search hard for partners, but silkworm moths don't have far to go. Silk farmers put the male and female moths together so that they can find mates.

Even though the domesticated silkworm moth doesn't have to search for a partner, it uses the same method of attracting a mate as wild moths do. Female moths produce a chemical with a powerful odor that male moths find very attractive. The photographs on these two pages show how a female's odor affects male moths.

8

The large female moth inside the box is producing a strong odor that has attracted the male moths. They are fluttering their wings so violently that some of the tiny scales are coming off.

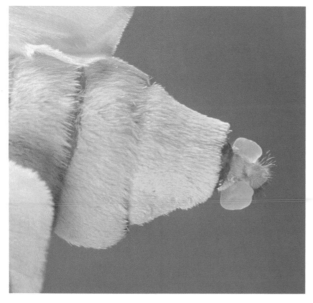

This yellow gland on the female moth's abdomen produces the chemical substance that attracts male moths.

The odor produced by a female moth comes from a special organ or gland located on her abdomen. This gland releases chemical substances known as **pheromones**. Insects and many other animals communicate with members of their own species by releasing pheromones. These chemicals can send different kinds of messages. Some warn of danger, while others tell animals to stay away from each other's territories. One of the most common uses of pheromones is to attract a mate.

A male moth does not have a nose to smell a female's pheromones. But he does have very large, sensitive antennae that pick up the chemicals released into the air. These chemicals tell him that a female of his species is nearby and that she is ready to mate.

A male moth's long, feathery antennae react to the chemicals produced by the female.

Left: A pair of moths mating. The female is the larger of the two. *Opposite:* A female moth laying her eggs on a cocoon

Silkworm moths mate by putting the tips of their abdomens together. **Sperm** cells pass from the male's body to the female's body, where they are stored in a kind of sac inside her abdomen. The sperm will eventually unite with, or fertilize, the female's **eggs** to produce new silkworm moths.

After a male moth has mated, his job is finished and he soon dies. But the female moth still has important work to do before her life is over. She must lay the hundreds of eggs that are inside her body.

The female's eggs develop in two organs called **ovaries**. The eggs leave the ovaries through tubes called **oviducts** and pass out of the female's body through an opening at the end of her abdomen. As the eggs move through the oviducts, they are fertilized by the sperm stored in the female's body.

A female silkworm moth usually lays about 500 yellow eggs that are no bigger than the head of a pin. The eggs have a glue-like covering that makes them stick to anything they touch. These tiny, sticky yellow eggs are the first stage in the metamorphosis of the silkworm moth.

The second stage in this remarkable process of growth is the larval stage. When a moth egg hatches, the tiny, worm-like creature that emerges is called a **larva**. Moth and butterfly larvae are also known as **caterpillars**, but the larvae of silk-producing moths have their own special name. They are called **silkworms**.

Silkworms hatch by eating a hole in the soft covering of their eggs. The newly hatched caterpillars are less than ⅛ inch (3 millimeters) long, with large black heads and bodies covered with hair. They look very different from the adult moths with their white wings and bodies.

A grove of mulberry trees

Eggs laid by wild moths and butterflies hatch according to natural cycles affected by weather conditions and other factors. The development of silkworm moth eggs is planned and controlled by the silk farmers.

After the eggs are laid, they are stored in refrigerators so that they do not start to develop. When the farmers want the eggs to hatch, they put them in **incubators**, machines that warm the eggs to the proper temperature for hatching. After about 20 days, the silkworms emerge from the eggs.

How do the silk farmers decide when the silkworm eggs should hatch? Their schedule depends on the timing of another natural process, the growth of mulberry trees. The bright green leaves of the mulberry tree are the natural food of silkworms. In Japan, mulberry trees grow new leaves in spring and keep their leaves until late autumn. During this period, Japanese silk farmers usually hatch out four to six groups of silkworms and feed them on fresh mulberry leaves.

15

Opposite: Workers on a Japanese silk farm feeding silkworms with mulberry leaves. Japan is the largest producer of silk in the world today. *Right:* Thousands of silkworms on a feeding table

Like silkworms, most moth and butterfly caterpillars have a special plant food that they eat. The females of different wild species lay their eggs on or near the plants that the caterpillars will need as food. But silkworm eggs are laid in farms and factories, far from the mulberry trees on which silkworms feed. The caterpillars that hatch from the eggs have to depend on people to feed them.

Silkworms have been changed so much by domestication that even if they were turned loose in a grove of mulberry trees, they probably could not feed themselves. Their ability to walk has become so weak that they cannot even move 12 inches (30 centimeters) to reach their food. On a silk farm, the newly hatched silkworms are put on feeding tables among heaps of chopped mulberry leaves. Even with their weak legs, the silkworms can get a good meal here.

Left: A silkworm just after hatching is in its first instar. *Right:* After its first molt, a silkworm enters its second instar. *Opposite:* In this photograph, the silkworm on the left is a third-instar silkworm; the one behind it is a second instar.

Kept supplied with mulberry leaves, a newly hatched silkworm eats almost without stopping. At this stage in its life, the silkworm's only job is to eat and to grow. Like other caterpillars, silkworms grow in a special way. Their soft bodies are covered by a skin that does not stretch as the caterpillars get bigger. Because of this, silkworms must develop new, larger skins from time to time. The process of growing a new skin and getting rid of an old one is called **molting**.

In their larval form, silkworms molt four times. The periods before or after the different molts, when the silkworms are eating and growing, are called **instars**. There are five instars in a silkworm's development. The pictures on the next few pages show silkworms in these five stages of their larval growth.

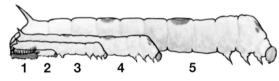

This drawing shows the actual size of a silkworm in its five instars.

1 2 3 4 5

INSTARS	1st	2nd	3rd	4th	5th
Days of eating and growth	2.5	2.5	3	4	8
Days of inactivity before molting	1	1	1.5	2	

This chart lists the number of days that a silkworm spends in each instar. At the end of the fifth instar, the silkworm changes from a larva to a pupa, the next stage in its metamorphosis.

This fourth-instar silkworm is getting ready to molt.

When a silkworm is ready to molt, it stops eating. It remains motionless, its head raised in the air, for at least one day before each molt. Japanese silk farmers say that the silkworm is "sleeping" during these periods of inactivity.

When the silkworm is asleep, changes are taking place inside its body. A new skin is forming underneath the old one. At the same time, a special liquid produced by the silkworm's body is dissolving the inner layers of the old skin, making it much thinner. When the new skin is completely developed, the silkworm wakes up and starts moving restlessly. It is time to shed the old skin.

The first thing that the silkworm does is to tighten its muscles and take in a lot of air so that its body swells. The increased pressure on the old skin causes it to split.

Holding on to the mulberry leaves with its little clawed legs, the silkworm begins to wiggle its way out of the old skin.

The silkworm is almost free of its old skin, revealing the brand new one underneath. It will soon begin eating again, and its first meal will probably be the skin it has just shed.

A silkworm that has just finished its fourth molt. Its new skin is loose and wrinkled.

After its fourth molt, the silkworm eats more than ever before. Its new skin is loose and wrinkled, but it will soon become smooth and tight as the silkworm stuffs itself with mulberry leaves. The silkworm is now in its fifth and final instar. This period of eating and growth is longer than each of the previous instars. During the fifth instar, the silkworm will eat 80 percent of all the food it consumes during its larval stage. At the end of this period, the silkworm will weigh over 10,000 times as much as it did when it hatched from the egg.

A fifth-instar silkworm eating a mulberry leaf. Silk farmers say that when large numbers of silkworms are munching on leaves, they make a sound like falling rain.

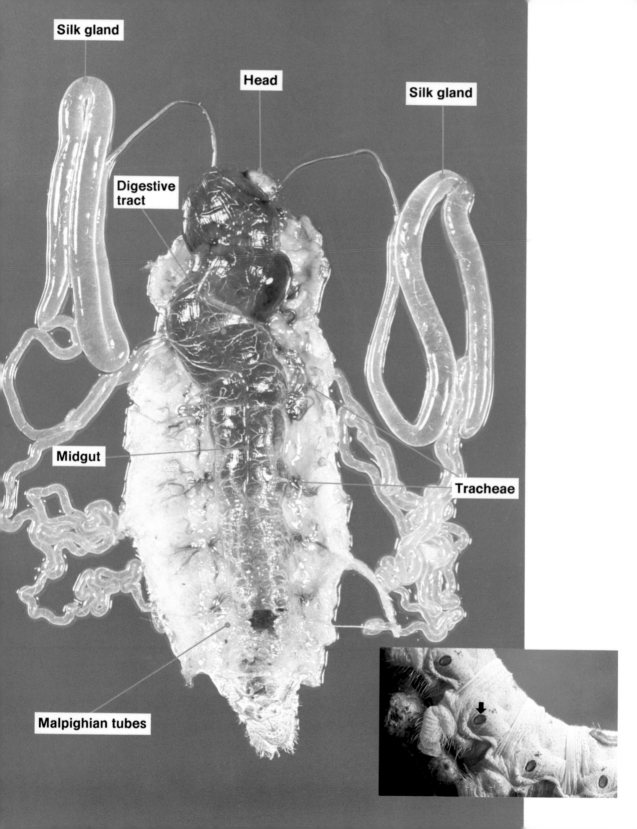

Silk gland

Head

Silk gland

Digestive
tract

Midgut

Tracheae

Malpighian tubes

The pictures on these two pages show some of the parts of a fifth instar's body. On its head, a silkworm has 12 **simple eyes** arranged in groups. These eyes can tell the difference between light and dark, but they cannot see images. The silkworm sends out its silk through a tube called the **spinneret**, located near the lower lip.

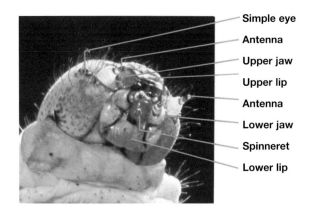

Simple eye

Antenna

Upper jaw

Upper lip

Antenna

Lower jaw

Spinneret

Lower lip

On the opposite page is a photograph of a dissected fifth-instar silkworm—one that has been cut open so that you can see inside. Below the silkworm's head is the large digestive tract, full of green mulberry leaves. In the lower part of the body are the yellow **Malpighian tubes**, which are part of the system that eliminates waste material. The **tracheae** make up the silkworm's respiratory system; these tubes are connected to the **spiracles**, or breathing holes, located on the sides of its body. (The small picture at the bottom shows some of the silkworm's 18 spiracles.)

Two of the most important parts of a fifth instar's body are the **silk glands**, the organs that produce silk. In this picture, they have been pulled out so that you can see how large and long they are.

This cross-section photograph of a silk thread was taken through a microscope. It shows the separate pieces of silk that are joined together to make a single thread.

During the silkworm's five instars, the silk glands have been growing larger. By the end of the fifth instar, they make up more than 25 percent of the silkworm's body weight.

Although the silk glands are not fully developed until the end of the fifth instar, they have been working during the earlier periods of the silkworm's life. Like most moth and butterfly caterpillars, silkworms constantly send out tiny threads of silk as they move along. These threads help the caterpillars to cling to leaves and other slippery surfaces. When a silkworm molts, it often uses silk threads to fasten the old skin to leaves or twigs while it wiggles out.

A silkworm's spinneret sends out a single thread of silk made up of two joined threads, one from each silk gland. The two threads are glued together by a sticky substance called **sericin**. Silk is in a liquid form inside the silkworm's body, but when it comes into contact with the air, it becomes hard.

This silkworm was photographed through a piece of glass. You can see how the caterpillar is using silk threads to get a foothold on the slippery surface. The figure-eight pattern of thread can also be seen.

As a silkworm sends out silk, it moves its head in a figure-eight pattern. The silk thread falls into this same pattern as it comes out of the spinneret.

The oakworm moth (above) is one of the wild moths that spin silk cocoons. Other moths are protected only by hard shells as they change into adults. Butterflies also use protective shells at this stage of their development.

The silk produced by silkworms and other caterpillars has one use that is more important than all the others. Silk is the material out of which cocoons are constructed. Many caterpillars make these protective coverings when they are getting ready to change into their adult forms. At this time, caterpillars are unable to move or defend themselves. The must have some way of hiding or of covering themselves so that they will not be attacked by enemies.

The domestic silkworm has no enemies, but it makes a protective cocoon just like its wild relatives. At the end of the fifth instar, silkworms stop eating for the last time.

Silkworms preparing to make their cocoons. Each silkworm occupies one space in the cardboard cocoon frame.

Silk farmers know that caterpillars are now ready to start their cocoons, so they put them on cardboard cocoon frames that provide support. Wild moths usually attach their cocoons to twigs and leaves.

A silkworm making support threads for its cocoon

Once the silkworms have been put into the cocoon frames, they go right to work. The first thing that they do is to send out short threads that become attached to the walls of the frame around them. These threads will form a support that the silkworms use while they are building their cocoons.

A silkworm's cocoon is made of a single unbroken thread of silk more than one mile (1.6 kilometers) long. In constructing the cocoon, the silkworm moves its head in the figure-eight pattern rapidly, as often as once every second. Silk comes out of the spinneret in tiny figure eights that overlap to form a thick, dense surface. The gummy coating of sericin on the silk thread causes it to stick together. After making the outer layer of the cocoon, the silkworm continues to add other layers from the inside.

The silkworm at the top has spent about 6 hours producing the support threads for its cocoon. The cocoon in the lower part of the frame has begun to take shape after 15 hours of work.

Left: An adult wasp and wasp larvae in a nest. Like silkworms, the larvae encase themselves in silk thread before changing into their adult forms. *Opposite:* Silkworm cocoons in many different stages of development

It takes at least two days for a silkworm to finish making its cocoon. During that time, it works constantly, spinning out silk and building up layer after layer. When the job is done, the silkworm is completely enclosed in a thick, strong shell of silk thread.

Many wild insects build sturdy silk cocoons, but the silk threads they use are not as long or as strong as those of the silkworm. For hundreds of years, silk farmers have done everything they could to improve the quality of the silkworm's silk. They have made sure that only the healthiest adult moths mated and reproduced. They have checked eggs and caterpillars carefully and destroyed those that seemed to be damaged in any way. As a result of these centuries of human effort, domestic silkworms now make only the best kind of silk.

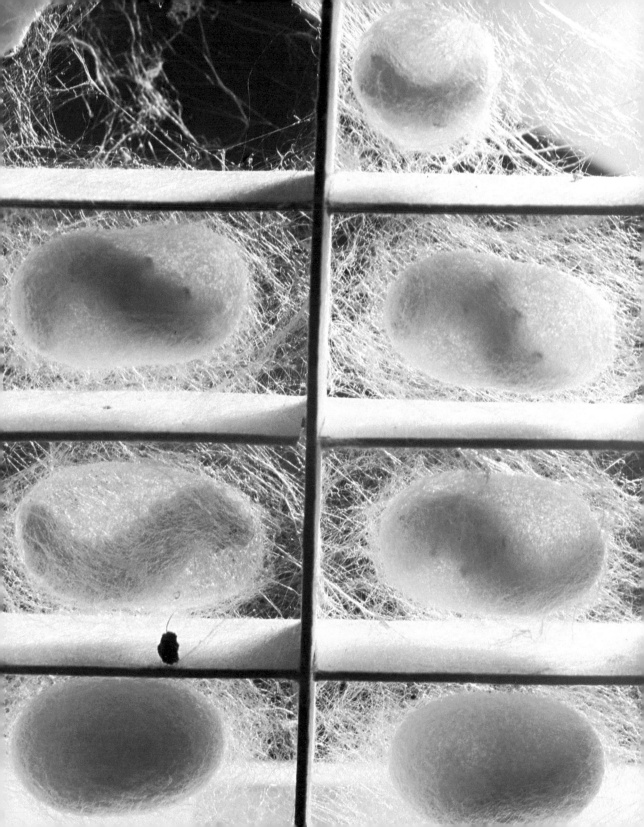

Opposite: This cocoon has been cut in half so that you can see the various stages in the development of a pupa. As the larval skin is pushed off, the soft yellowish body of the pupa is revealed. After the transformation is complete, the covering of the pupa's body will become hard and brown.

Of course, silkworms do not know that they are making high-quality silk for human beings. They are doing only what their instincts tell them to do at this stage in their lives. Building silk cocoons is part of the natural process of development that will eventually produce adult silkworm moths.

Within the protection of its cocoon, a silkworm enters the third stage of metamorphosis. It changes from a larva to a **pupa.** This change begins when the silkworm sheds its skin for yet another time. The old skin tears and is pushed back just as in earlier molts. But what is underneath is not a larva with a new skin; it is a soft yellowish body that has a whole different shape. The silkworm has taken on its pupal form. Now its body will begin to go through the great changes that will transform it into an adult moth.

Most silkworms raised on silk farms are not allowed to develop into adults. When an adult moth comes out of its cocoon, it makes a large hole that breaks the long silk thread. To prevent the silk from being damaged, silk farmers kill most of the pupae before they become adults.

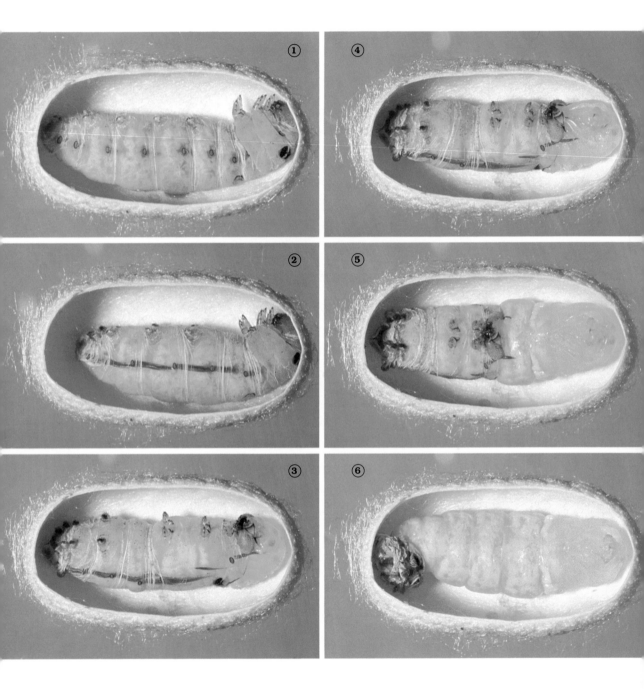

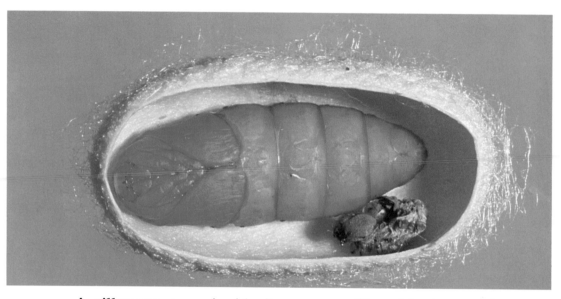

A silkworm pupa inside its cocoon. The object near the right end of the cocoon is the larval skin that the silkworm shed when it became a pupa.

The farmers kill the pupae by putting the cocoons in a hot oven. After the pupae are dead, the cocoons are sent to a special factory where the silk thread is unwound. The cocoons are first placed in hot water to dissolve the sticky sericin that holds them together. The thread from a single cocoon is too fine to be handled by itself, so threads from as many as 10 cocoons are put together and wound onto a reel. Because the threads are still coated with the dissolved sericin, they stick together to form one large, strong thread. After being treated in other ways, this thread will be used to make soft, lustrous silk fabric.

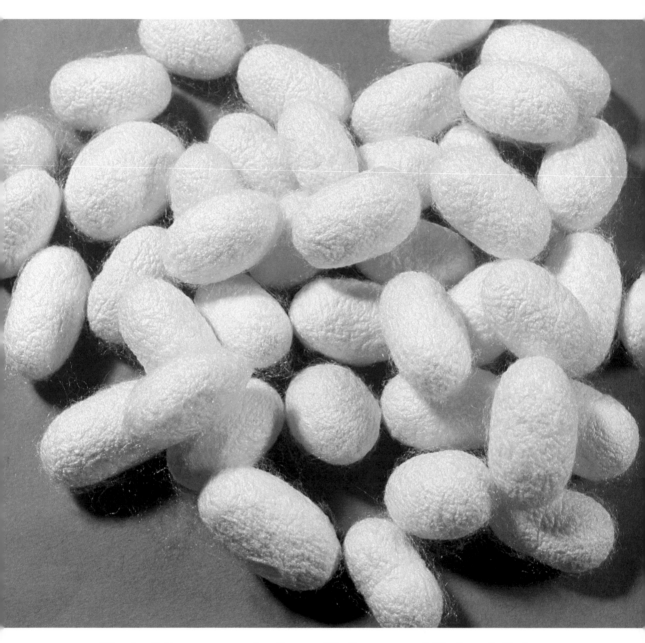

Most of the pupae raised on silk farms are killed and their cocoons used for silk thread. A few are allowed to continue their development and become adult moths.

If a pupa is not destroyed inside its cocoon, it will continue its natural development. About three weeks after it formed inside the cocoon, the pupa will have become an adult moth, and metamorphosis will be complete.

The pupal stage is in many ways the most remarkable of the four stages in the silkworm moth's life. During this time, the body of the silkworm changes into the very different body of an adult moth. Inside the hard pupal shell, the larva's short little legs are replaced by the long, jointed legs of the adult. Its thin, worm-like body is transformed into a short, thick body covered with white scales. An adult's long, feathery antennae replace the tiny sense organs on the silkworm's head. Large compound eyes and scaly white wings develop.

Inside the cocoon, the pupa remains motionless while these tremendous changes take place. When the development is complete, the hard pupal covering splits open and the adult moth emerges.

The pictures on the opposite page show an adult coming out of the pupal shell. In the top picture, the moth's head has just emerged, and you can see its big black eyes. The middle picture was taken two minutes later; by now the moth has pulled out its antennae and two pairs of legs. Seven minutes later (bottom picture), its wings have begun to emerge.

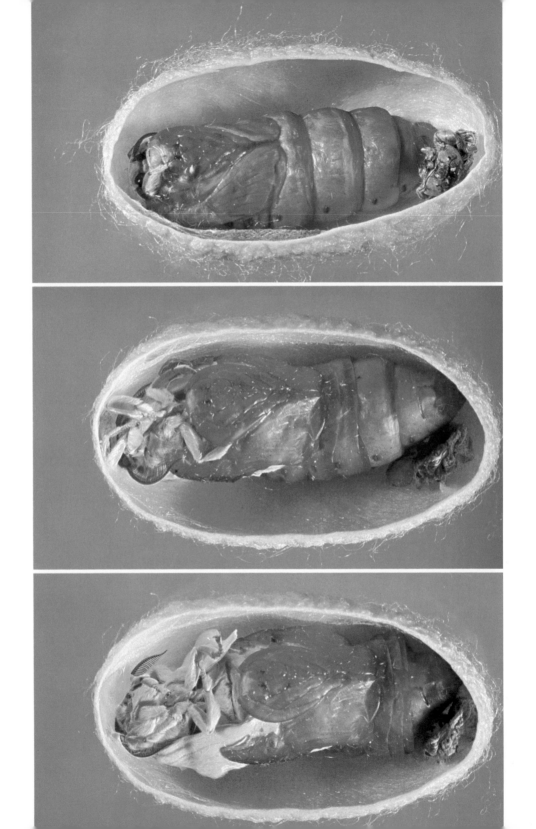

Above: The top of the moth's head pushes through the hole in the cocoon. *Right:* The moth has its upper body out. You can see drops of the dissolving liquid on its head and leg.

A moth works hard to get out of the pupal shell, but it has an even bigger job ahead of it. It must break out of the thick silk cocoon that provided protection during its pupal development.

The moth cannot bite a hole in the covering around it as the larva does when it comes out of the egg. Like other adult moths and butterflies, the silkworm moth has no teeth or biting parts in its mouth. To make a hole in the cocoon, the moth must use a special liquid that its body starts producing when it is ready to emerge. This liquid comes out of the insect's mouth. It softens and dissolves the silk so that the moth can push its way though the cocoon.

The pictures on these pages show a moth gradually emerging through the hole it has made in the cocoon.

As the moth continues to work its way out of the cocoon, its crumpled wings begin to emerge.

It takes about 15 minutes for a silkworm moth's wings to open completely.

After a moth has gotten all six of its legs out of the cocoon, it uses them to get the rest of its body out. Bracing its legs against the side of the cocoon, it struggles to pull out its heavy lower body. (In the picture on the opposite page, you can see a moth using its legs in this way.)

Finally the silkworm moth has escaped from the cocoon. As the final step in its development, it must now extend its wings, which are soft and crumpled after being folded inside the pupal shell. As the moth clings to its cocoon, blood is pumped into its wings, gradually expanding and hardening them. But even after the moth's wings have opened completely, they look small and weak compared to its heavy body.

In fact, the wings of a silkworm moth are small in proportion to its body. Unlike wild moths and butterflies, this domesticated insect has very little need for wings at all.

Racks of cocoon frames on a silk farm. Each year, silk farmers take millions of yards of silk thread from the cocoons of silkworm moths.

It does not have to fly in order to find a mate or to escape from an enemy. If silkworm moths had large strong wings capable of real flight, they would cause problems for silk farmers. Over the centuries, farmers have encouraged the development of weak-winged moths by allowing only moths with that characteristic to reproduce. Today silkworm moths can hardly fly at all. They can only flutter their wings and hop from place to place.

After coming out of its cocoon, a wild moth flies away on its newly opened wings. A silkworm moth remains on or near its cocoon, waiting for the human help it must have in order to continue the natural cycle of its life.

GLOSSARY

antennae (an-TEN-ee)—sense organs on the heads of insects that respond to touch and odor. The singular form of the word is *antenna*.

caterpillar—the larva of moths and butterflies

cocoon (kuh-KOON)—a covering composed partly or entirely of silk that serves as protection for an insect during its pupal development. Moths, bees, and wasps are some of the insects that make cocoons.

domesticated—tamed and cared for by human beings. Chickens, cows, pigs, and silkworm moths are all domesticated animals.

egg—the female sex cell, which can unite with a male sex cell to produce new life

incubator—a machine that keeps eggs at the proper temperature for hatching

instars—the periods between a larva's molts, during which it eats and grows. A larva going through these different periods is sometimes referred to as a first instar, a second instar, etc.

larva—the second stage in complete metamorphosis. A larva is a worm-like creature that looks very different from the adult insect. The plural form of the word is larvae, pronounced LAR-vee.

Malpighian tubes (mal-PIG-ee-un)—tubes in an insect's body that collect and eliminate waste material

metamorphosis (met-uh-MOR-fuh-sis)—the process of growth and change that produces most adult insects. Moths, butterflies, bees, and ants go through a four-stage development known as complete metamorphosis; the four stages are egg, larva, pupa, and adult. Another process of development called incomplete metamorphosis has only three stages: egg, nymph, and adult. Grasshoppers, crickets, and cockroaches develop in this way.

46

molting—the process of growing a new skin and shedding the old one

ovaries—organs in a female's body that produce eggs

oviducts—tubes that carry a female insect's eggs out of the ovaries and into the chamber from which they leave the body

pheromones (FER-uh-mohns)—chemical substances produced by an animal's body that are used to communicate with other animals of the same species

pupa—the third stage in complete metamorphosis. During the pupal stage, the body of the adult insect is formed. The plural form of the word is pupae, pronounced PEW-pee.

sericin (SER-uh-sun)—a sticky substance produced by a silkworm's body that holds strands of silk thread together

silk glands—organs in a silkworm's body that produce liquid silk

silkworm—the larva of the silkworm moth, Bombyx mori

simple eye—an insect eye capable of telling the difference between light and dark. Larvae have only simple eyes, while many adult insects have both simple and compound eyes. Compound eyes can see images and colors.

sperm—the male sex cell, which can unite with a female sex cell to produce new life

spinneret—the tube on a silkworm's head through which silk is expelled

spiracles (SPEAR-uh-kuhls)—breathing holes located on the sides of an insect's body

tracheae (TRAY-kee-ee)—tubes that carry oxygen throughout an insect's body. The singular form of the word is trachea, pronounced TRAY-kee-uh.

INDEX

adult moth, 7, 8; emergence of, 38, 40, 43

antennae, 6, 10, 11, 38

butterflies, 6, 14, 15, 26, 28, 40, 44

caterpillars, 14, 17, 18, 26, 28

cocoons, 5, 28, 29, 34, 40; making of, 30-32; treatment of, on silk farms, 36, 37

domestication, 6, 8, 17, 32, 45

eggs, 12, 17; hatching of, 14-15

eyes, 25, 38

fertilization, 12

frame, cocoon, 5, 29-30, 31, 45

incubator, 15

instars, 18-23, 25, 26, 28

Japan, silk farming in, 6, 15, 17, 20

larva, 14-32, 34, 38, 40

Malpighian tubes, 25

mating, 8, 12, 45

metamorphosis, stages in: egg, 12; larva, 14-32; pupa, 34-38; adult, 7, 8, 38-43

molting, 18, 20-21, 22

moths: characteristics of, 6, 14, 26, 40; living in the wild, 8, 15, 17, 28, 29, 32, 44

mulberry leaves as food for silkworms, 15, 17, 18, 22, 23, 25

ovaries, 12

oviducts, 12

pheromones, 10

pupa, 34-38

sericin, 26, 30, 36

sex cells, 12

silk: fabric made of, 5, 36; production of, by silkworms, 25-28, 30, 32; treatment of, on silk farms, 32, 26

silk farming, 6, 15, 17, 23, 29, 32, 34, 36, 45

silk glands, 25, 26

simple eyes, 25

sperm, 12

spinneret, 25, 26, 27, 30

spiracles, 25

tracheae, 25

wings of adult moth, 6, 7, 44-45; extending of, 41, 43